TURTLES
HEADING HOME!

Liza Ketchum, Phyllis Root, and Jacqueline Briggs Martin

Charlesbridge

INTRODUCTION

Sand quivers at dawn on a warm beach along the Gulf of Mexico.

Beneath the sand, baby Kemp's ridley turtles—small as the palm of your hand—peck at their eggshells and emerge into a nest alive with squirming hatchlings.

Climbing over one another, a frenzy of baby turtles scrambles across the sand.

Drawn to light over the ocean, they rush to escape hungry predators: seagulls, foxes, vultures, dogs, and raccoons.

Turtle hatchlings rush toward the ocean.

The baby ridleys have never seen ocean surf, but when the first waves break over them, they know what to do. They paddle, dive, and swim for their lives.

These turtles have begun a long journey that will take them from the Gulf of Mexico to the Sargasso Sea in the Atlantic Ocean.

A few years later—if they survive predators, shrimp nets, and the blades of motorboat engines—they will swim north to the warm waters of the Gulf of Maine and Cape Cod Bay. There they will feast on their favorite food: crabs.

1: RESCUE

A wave sweeps the beach. Retreats.

Another wave comes.

The December beach is cold, dark, empty. But wait—what did the wave leave behind?

A sea turtle, big as a dinner plate. It doesn't move. The turtle is stranded, far from home.

Like other reptiles, turtles are ectothermic. Their bodies take on the temperature of the surrounding waters: This turtle is cold stunned.

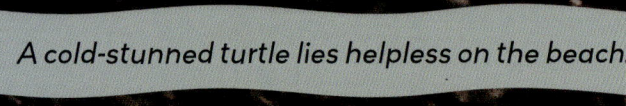
A cold-stunned turtle lies helpless on the beach.

Climate change has warmed the waters in the Gulf of Maine and around Cape Cod, which used to cool earlier in the fall, warning the turtles it was time to swim south.

Now the sea stays warmer longer, fooling the turtles into staying later.

MASSACHUSETTS

RHODE ISLAND

Gulf of Maine

Cape Cod Bay

Cape Cod

Atlantic Ocean

Martha's Vineyard

Nantucket

As turtles try to swim south from the Gulf of Maine, they may get stuck inside the hook that arcs around Cape Cod Bay.

A cone of light rakes across the sand.

"Found one," a volunteer calls softly.

Her name is Heather Pilchard, and she often walks along the beach at night, searching for turtles.

A second beam sweeps over the turtle's ridged shell.

"Is it alive?" her patrol partner, Tina Maloney, whispers.

"It's not moving. Looks like a ridley," responds Heather.

"I'll cover it." Tina crouches down. "You get the sled."

A volunteer spots a cold-stunned turtle.

Volunteers use strong flashlights as they search the beach for turtles.

Heather and Tina belong to a large group of volunteers who care deeply about sea turtles. More than two hundred of them take turns patrolling the beaches during winter—twice a day at high tide, when turtles might be washed ashore. Because ridley turtles are an endangered species, volunteers must have federal permission to move them from the beach.

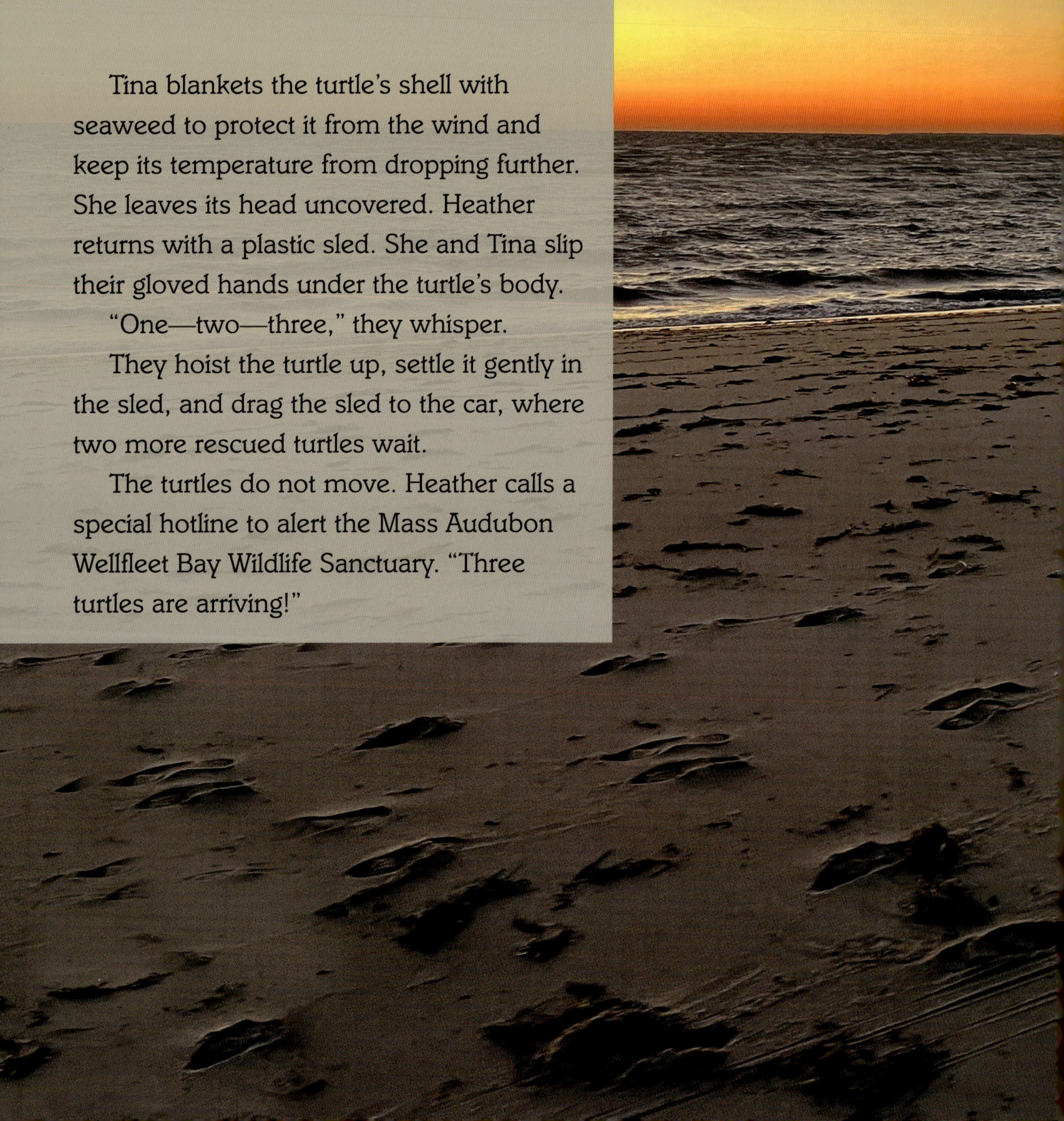

Tina blankets the turtle's shell with seaweed to protect it from the wind and keep its temperature from dropping further. She leaves its head uncovered. Heather returns with a plastic sled. She and Tina slip their gloved hands under the turtle's body.

"One—two—three," they whisper.

They hoist the turtle up, settle it gently in the sled, and drag the sled to the car, where two more rescued turtles wait.

The turtles do not move. Heather calls a special hotline to alert the Mass Audubon Wellfleet Bay Wildlife Sanctuary. "Three turtles are arriving!"

Like Heather and Tina, Sasha Milsky sometimes uses a sled to transport turtles along the beach.

Rescued Kemp's ridleys are nestled into banana boxes and brought indoors.

The sun rises as the car arrives at the sanctuary. Scientists and volunteers, bundled against the cold, wait next to a line of banana boxes—just the right size for Kemp's ridleys.

A few boxes already hold turtles, snuggled onto beach towels. The new arrivals soon fill other boxes. Almost all of the rescued turtles are Kemp's ridleys, the world's most endangered sea turtle.

Volunteers carry the banana boxes inside. No one speaks. They set the ridleys on long tables in a clean, quiet room next to boxes of other rescued turtles.

The room is cold—only 55 degrees Fahrenheit. The turtles need to warm slowly, or they will die.

Karen Dourdeville, sea turtle research coordinator, and Sasha Milsky, sea turtle rescue technician, inspect the new arrivals. They speak in low voices so they don't scare the traumatized turtles.

Sasha lifts a ridley onto the counter and sets it on a clean towel. She and Karen measure its length, width, and weight and record their findings. The turtle doesn't stir. They dip a cotton ball in clear water and dribble it over the turtle's eyes to clean off the sand.

The turtle still doesn't move.

Sasha Milsky uses calipers to measure a turtle. Mass Audubon Wellfleet Bay collects data for turtle research and to help with treatment.

A turtle rests on a towel during examination. Every turtle is handled gently.

"What's its number?" Karen asks.

"274," Sasha says.

That number shows how many turtles have stranded on Cape Cod beaches so far this year. By the end of the winter, the number will be close to eight hundred. Not all will survive.

"Is it dead?" Sasha asks.

Karen gently moves the turtle's flipper. The flipper jerks.

"It's alive!"

2: REHAB

Karen and Sasha work quickly to assess the other turtles. A car waits to rush the ridleys to the New England Aquarium's Sea Turtle Hospital, ninety miles away. Inside the car it is also 55 degrees.

Ten minutes before reaching the hospital, the volunteer driver calls ahead. "Get ready. Turtles arriving!"

When the driver rings the doorbell, staff biologists, interns, and volunteers hurry to carry in the banana boxes of turtles.

Sometimes turtles travel in vans.

New England Aquarium
Marine Animal Rescue Team

Dylan Marat loads a turtle into Kathy Herrick's car. Stunned, stranded turtles need medical treatment as soon as possible.

Inside, the turtle hospital is quiet.

No sirens. No phones beeping. No shouting.

Just soft voices and water lapping the sides of deep blue tanks—tanks already filled with Kemp's ridley turtles.

Turtles swimming. Turtles resting on the bottom. Turtles coming up for air.

A recovering turtle swims in a tank. The water temperature in the tank is carefully controlled.

Turtles share a tank as they become stronger.

The new arrivals are too sick for the tanks. Some are coated with sand and algae. Aquarium staff and volunteers wash each ridley gently. They check the number on its flipper tag, give the turtle a matching zip-tie "bracelet," and pass the turtle on to the aquatic veterinarian, Dr. Kathy Tuxbury.

Natashja Molina lifts a sand-covered turtle from a box. She will carefully wash it clean.

Dr. Kathy Tuxbury examines an injured turtle with Alyssa Kaufold's help.

Dr. Kathy listens to the turtle's heartbeat, checks its eyes, takes its temperature with a rectal thermometer, and inspects its body for cuts and abrasions. She also takes a radiograph (a kind of X-ray) to see if it has pneumonia, the biggest danger to rescued turtles.

The ridley might need medicine injected through a long needle into the folds of its neck. Dr. Kathy makes a plan to treat it that she hopes will save the turtle's life. Of the turtles treated at the hospital, as many as 85 percent of them will return to the ocean where they belong.

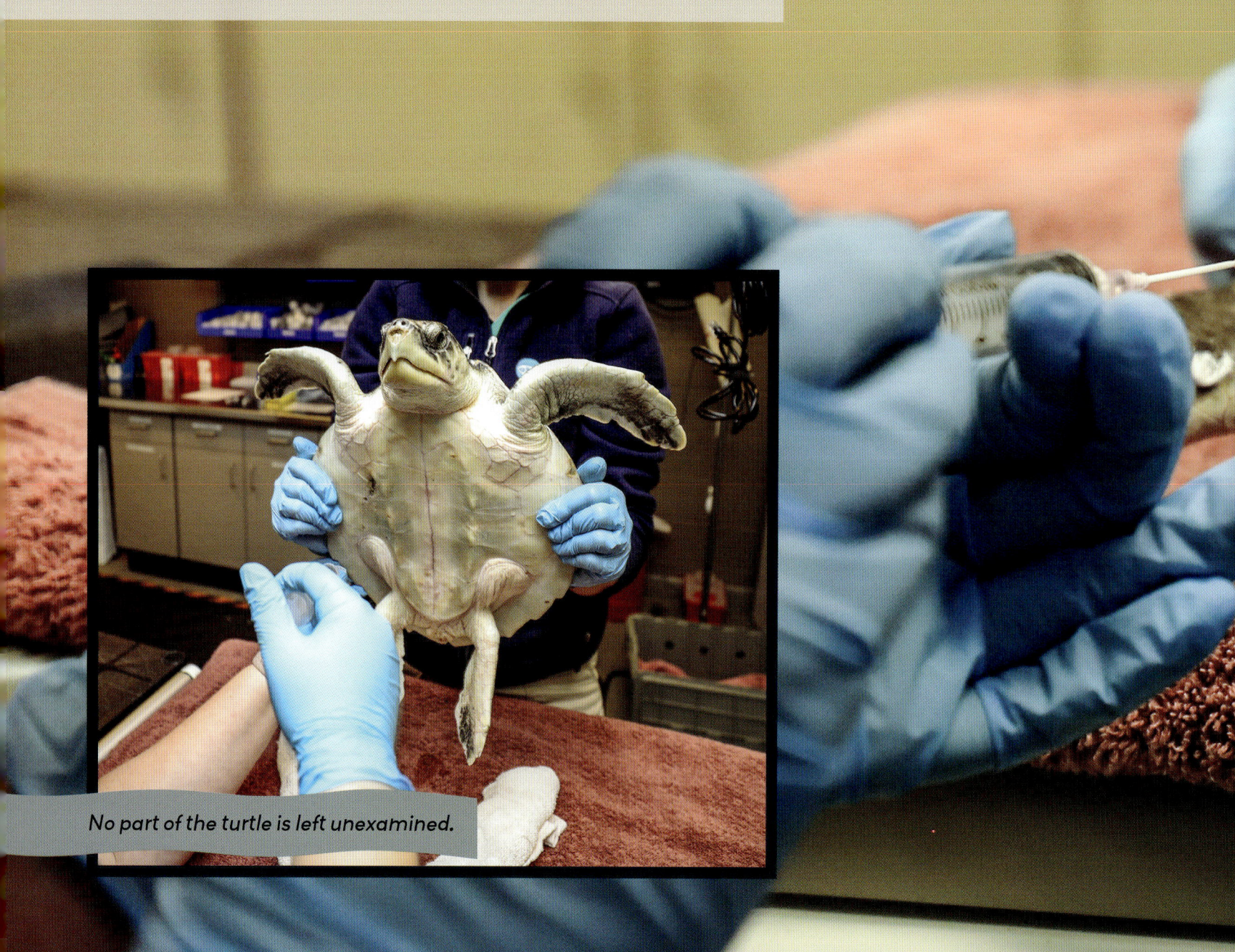

No part of the turtle is left unexamined.

Hospital staff inject a turtle with medicine.

After a turtle is evaluated and deemed ready, a staff member lowers it into a deep tank with water at 55 degrees—a little warmer than the turtle's cold-stunned body.

Turtles can't breathe underwater. Sometimes, an injured turtle might be too weak to lift its head to breathe. Intern Piper Maguire spots one ridley resting too long on the tank floor.

A turtle is lowered into water for the first time since it was stranded. Will the water feel like home?

Piper Maguire helps a turtle keep its head above water.

NE RESCUE 4 TUR

She scoops the turtle up, cups her hands under its body, and holds it at the surface until it sips the air.

She lets it sink back to the bottom and watches to make sure it doesn't drown. If the ridley doesn't swim, she might place it on a tiny "surfboard" to keep it afloat.

The first day, a turtle swims for only twenty minutes to an hour. If it's swimming well on the second day, it goes into a 65-degree tank. On the third day, it moves to a 75-degree tank, the warmest of all.

As the ridleys recover, they're fed soft food at first: mashed bok choy, herring without bones, and squid with the beak removed. Some turtles prefer squid, but it's not as nutritious as herring. Staff members fool these turtles by wrapping a piece of squid around a chunk of herring.

Staff members and volunteers work together to rehabilitate the turtles.

But here's the important rule: "No solid food until they've pooped!"

If there is no sign of poop in the tank water, veterinarians x-ray the turtle to see if food is moving through its intestinal tract.

Kemp's ridleys, like the green sea turtle shown here, love eating squid!

Strong, swimming turtles are ready to leave the hospital.

Even when a turtle is eating (and pooping) and swimming strongly, it still might not be ready to return to the ocean. But cold-stunned turtles keep arriving, and many need special care.

To make room, turtles that are doing better must move to another facility where they can continue to recover.

How will they travel?

A plane waits for its turtle passengers.

3: RELEASE

They fly!

Volunteer pilots from all over the country fly their small planes to Massachusetts to pick up turtles—loaded in banana boxes again—and transport them to the next stop on their journey.

They are part of an organization called Turtles Fly Too.

A New England Aquarium staff member helps pilot Matt Barnes load boxed turtles into his plane.

Pilot Ken Andrews flies from Michigan in his Mitsubishi Marquise plane. With the passenger seats removed, his plane can carry as many as sixty turtles. He stacks the boxes almost to the ceiling. He calls his plane a workhorse.

Often his wife, Kristi, is his copilot.

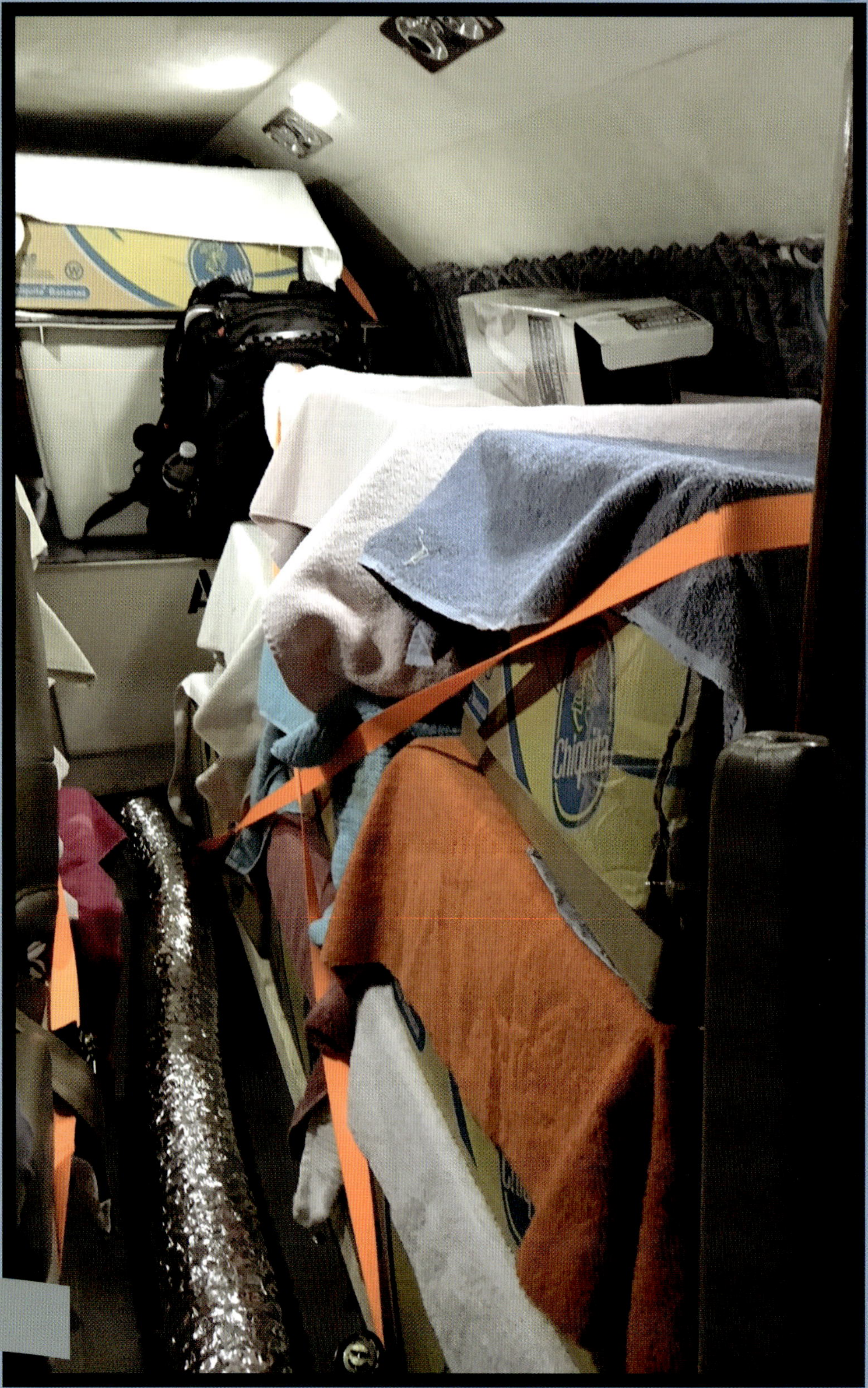

Turtles need safety belts, too.

Their young son is small enough for an important job: He crawls in back and checks a thermometer to make sure the air temperature stays at 70 degrees. If it's too cold, he snakes a hose full of warm air into the back, waving it above the boxes to heat the cargo area.

Pilot Ken Andrews and his son Christopher settle a turtle in the plane. People of all ages can help turtles.

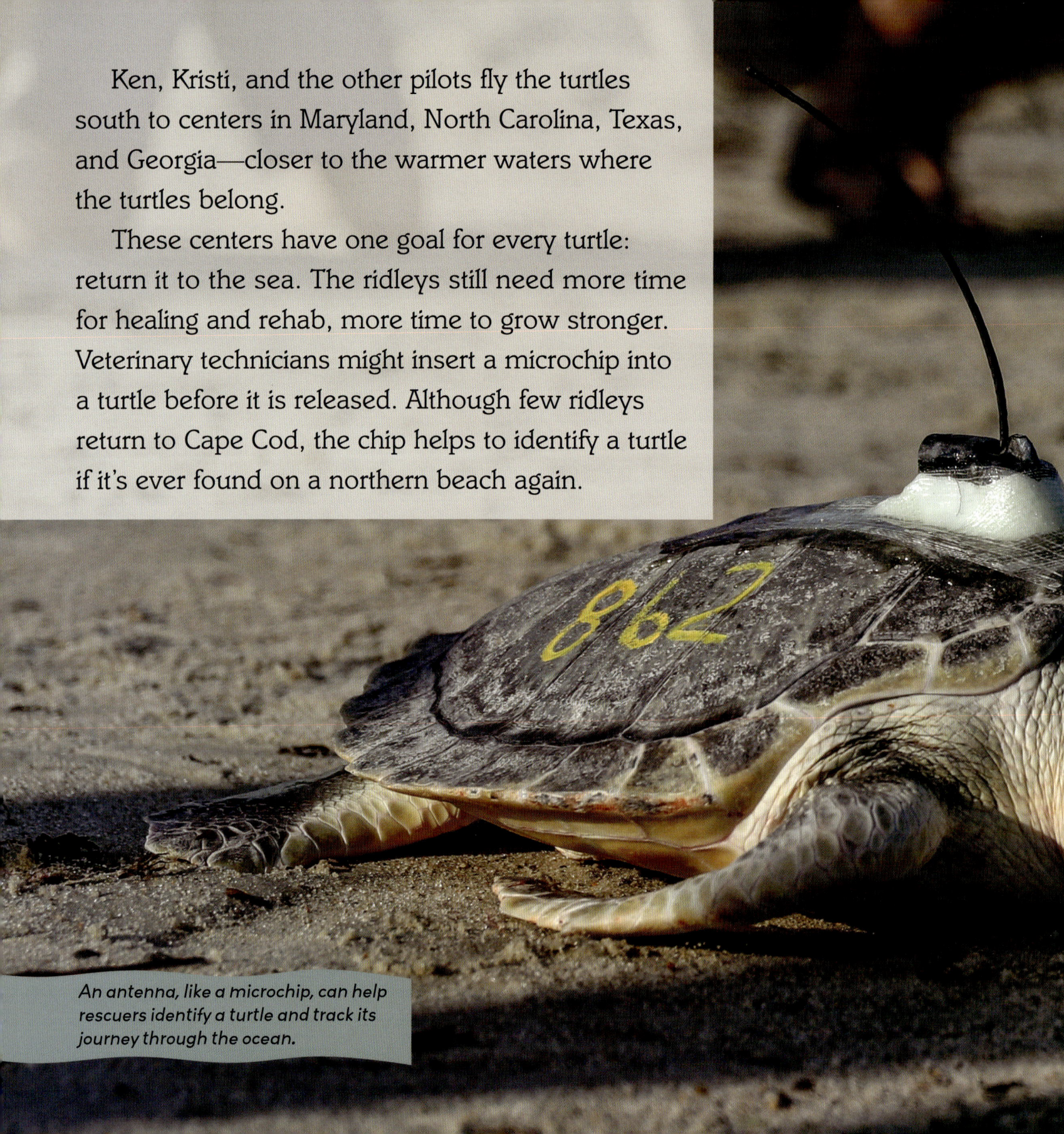

Ken, Kristi, and the other pilots fly the turtles south to centers in Maryland, North Carolina, Texas, and Georgia—closer to the warmer waters where the turtles belong.

These centers have one goal for every turtle: return it to the sea. The ridleys still need more time for healing and rehab, more time to grow stronger. Veterinary technicians might insert a microchip into a turtle before it is released. Although few ridleys return to Cape Cod, the chip helps to identify a turtle if it's ever found on a northern beach again.

An antenna, like a microchip, can help rescuers identify a turtle and track its journey through the ocean.

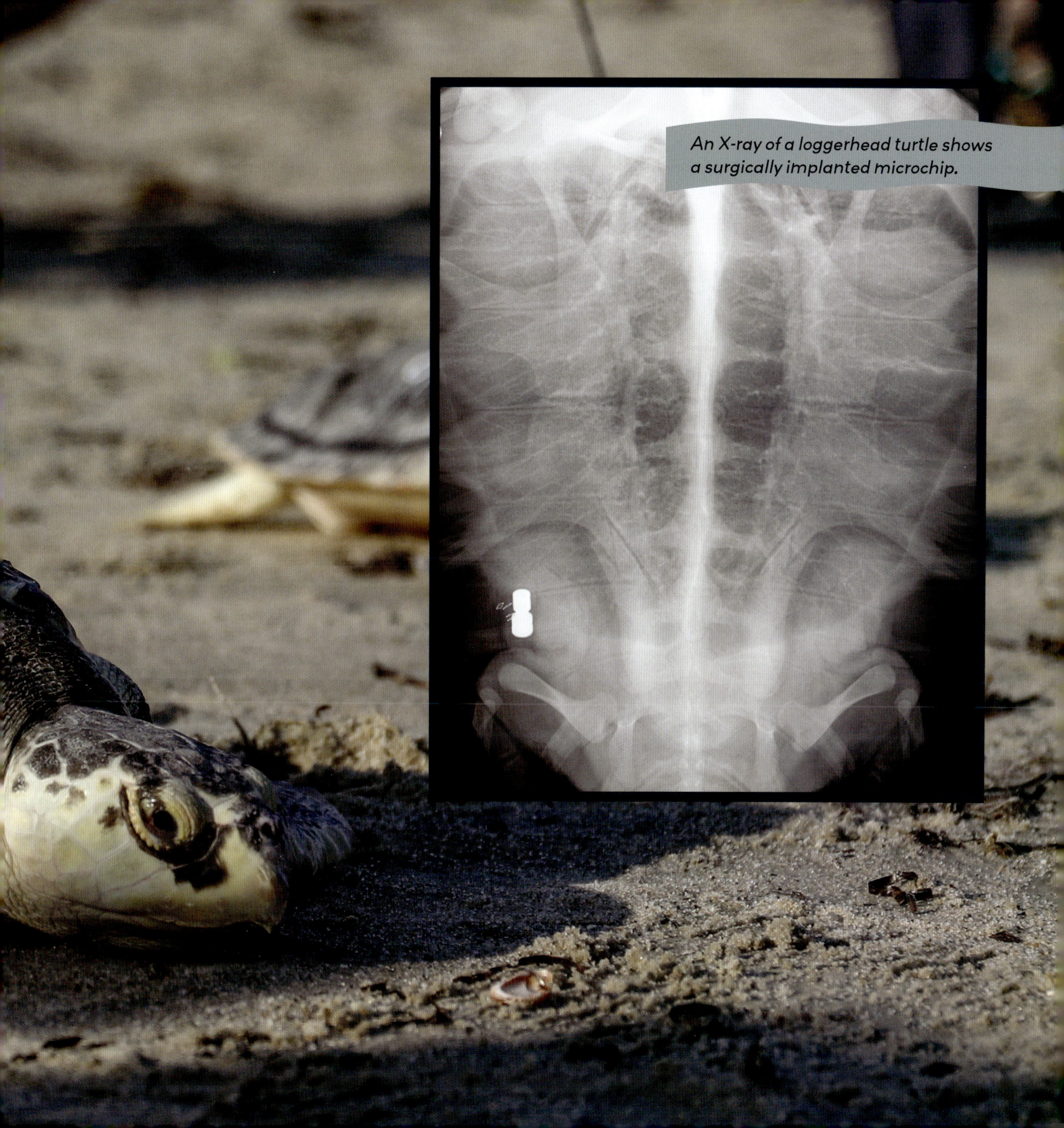

An X-ray of a loggerhead turtle shows a surgically implanted microchip.

Finally, on a warm, clear day, the turtles arrive in their boxes at a southern beach. Sometimes an announcement goes out: Turtle release today!

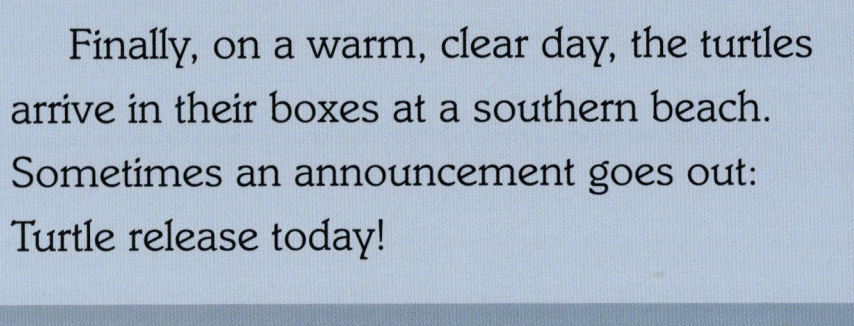

This turtle is eager to return to the sea.

People gather on the beach to watch. Carefully, staff members, volunteers, and sometimes children carry the turtles over the sand.

The ridleys flap the air, ready to swim.

Can they smell the ocean?

Do they hear the waves?

Staff and volunteers prepare to release turtles on a beach in Georgia.

Set down in shallow water, a ridley hauls itself forward in an awkward breaststroke. A wave washes over its shell—and retreats.

Another wave breaks, spreading foam. The ridley raises its head, looks around, and sips the air. It knows what to do. It strokes away from the shore—and dives.

A turtle makes its way back into the ocean.

Farther out, the turtle surfaces, takes another breath. Waves wash over its olive-brown shell and the beautiful cross-hatching on its flippers.

Then the ridley is gone.

Returned to the sea, at last.

Home.

ABOUT KEMP'S RIDLEY TURTLES

On a dry, blustery spring day, thousands of turtles gather in the sea, next to the same beach in the Gulf of Mexico where they hatched years before. How do they all know to arrive at the same time? Scientists are not sure.

After the turtles mate offshore, female turtles crawl onto the beach in an event called an arribada, *arrival* in Spanish. The females dig their nests in the sand, lay eggs, cover them, and "dance" on the nest to tamp the sand down before returning to the sea. Nesting is the only time healthy Kemp's ridley females go ashore.

Two months later the eggs hatch. Tiny turtles, each weighing about half an ounce (less than three nickels), dodge birds and other predators and make their perilous journey to the ocean. Healthy Kemp's ridley males will never go ashore again.

Some hatchlings are swept out into the ocean to the Sargasso Sea, an area bounded by currents where sargassum seaweed grows. Here they might spend many years eating tube worms, sargasso swimming crabs, jelly animals, and other tiny sea creatures that live in the seaweed. Some turtles make their way slowly up the Atlantic coast before returning to warmer waters. When Kemp's ridleys return to their nesting beach in twelve or thirteen years, they are fully grown. They weigh up to 100 pounds and measure about 2 feet in length, but they are still the smallest of Earth's sea turtles.

They are also the most endangered.

Like Kemp's ridleys, olive ridleys gather by the thousands to mate and lay their eggs.

In 1947 more than 42,000 Kemp's ridleys crawled ashore to nest. By 1985 the number of nesting female Kemp's ridleys had shrunk to fewer than 250. What happened?

People stole turtle eggs to sell or to eat. Turtles drowned when they were caught in the nets of shrimp trawlers and couldn't come up for air.

The numbers of nesting turtles rose but plunged again in 2010, partly because of the Deepwater Horizon oil spill in the Gulf of Mexico. Now Kemp's ridley nesting areas are protected, and shrimp trawler nets must be fitted with devices that allow captured turtles to swim free.

Kemp's ridleys rely on a single nesting site. So scientists have established a second site. They transported Kemp's ridley eggs from the arribada site and reburied them on Padre Island, Texas. Turtles that hatch on the island now return there to nest. In 2021, conservationists estimated that the total number of nesting Kemp's ridley females was between 7,000–9,000.

Today Kemp's ridleys face a new threat. Climate change warms the ocean and fools sea turtles. They remain in northern waters too late in the season. Of the six species of sea turtles that swim in the Atlantic Ocean, Kemp's ridleys are the ones most often cold stunned and stranded on Cape Cod.

WHO IS SAVING TURTLES?

Many professional staff—including aquatic veterinarians, biologists, and researchers—work to save cold-stunned turtles. Hundreds of volunteers are also essential to the turtle-rescue team.

Beach Walkers and Drivers

In 1974 Bob Prescott, a wildlife biologist and former director of the Mass Audubon Wellfleet Bay Wildlife Sanctuary, was walking on a winter beach. He spotted a cold-stunned Kemp's ridley sea turtle and wondered: "What's *that* doing here, so late in the year?" No one knew how to care for it. As more turtles stranded every winter, Bob and his staff created a volunteer program to treat injured turtles. Now 200 trained volunteers patrol the beaches of Cape Cod Bay, Martha's Vineyard, and Nantucket Island and rescue 800–1,200 cold-stunned turtles a year.

Finding turtles is hard work. Volunteers slog through wet sand, rain, and sometimes even snow. After dark, volunteers work in pairs for safety. (Night patrol can be dangerous because it's hard to see the waves.) Beach walkers keep plastic sleds or toboggans, rubber boots, winter jackets, old towels, and flashlights or headlamps in their cars.

Trained volunteer transport drivers are also on call, day and night, to move turtles from Cape Cod to the New England Aquarium's Sea Turtle Hospital.

Hospital Staff and Volunteers

The New England Aquarium has been helping save Kemp's ridleys for more than twenty-five years. In a single year they might care for more than five hundred turtles and release 85 percent of them back into the wild.

Full-time, part-time, and seasonal professionals, working under Director Adam Kennedy, care diligently for the turtles at the aquarium's sea turtle hospital.

Some volunteers are interns studying to be aquatic veterinarians or scientists. Others are students, retirees, or people who take a weekly day off from work. Becky Lash teaches entomology at the Woods Hole Children's School of Science. She helps out every Friday. So does Sarah Capozzoli, a vet tech who works in a veterinarian's office. She has been a volunteer for more than twenty years!

Turtle Fliers

Getting sick turtles to their destination quickly is critical to their survival. Pilots who fly with Turtles Fly Too transport turtles from Massachusetts to southern aquariums and other facilities. Known as "turtle fliers," about 450 pilots donate their planes and fuel. Pilot Ken Andrews takes time off from his day job to fly from Michigan to Massachusetts. From there he flies dozens of turtles to aquariums in Maryland, North Carolina, and Georgia— and even the Kennedy Space Center in Cape Canaveral, Florida! Volunteers at each airport telephone the pilots, coordinate their flights, check the weather, and monitor takeoffs and landings.

Why does everyone work so hard to save the turtles?

Ken Andrews says, "Flying is a privilege, and I want to give back."

Bob Prescott says, "Turtles are very charismatic, and they need our help. Stranding is not going away, but our program is a real success story."

Adam Kennedy puts it this way: "Our hope is that saving each individual sea turtle will give future generations the chance to appreciate these magnificent creatures."

Even if you don't live by the ocean, you can help turtles.

Volunteers and staff from the New England Aquarium and Mass Audubon Wellfleet Bay gather to celebrate a successful turtle release.

WORK TO FIGHT CLIMATE CHANGE

Burning fossil fuels, such as coal, natural gas, and oil, releases a gas called carbon dioxide. Too much carbon dioxide warms the earth and its oceans, causing climate change.

We know climate change tricks turtles into staying longer than they should. When the water suddenly turns cold—often overnight—turtles that haven't left may become cold stunned.

Climate change also warms the sandy beaches where turtles nest. Whether a turtle becomes male or female depends on the temperature of the sand where its egg is laid. Warmer sands create more eggs that will hatch into females; cooler sands produce more males. Turtle survival depends on equal numbers of male and female turtles.

As the earth warms, seas levels rise. Higher seas can flood turtle nests.

Here are some things you can do to help:
- Urge people to cut down on the use of fossil fuels.
- Encourage your friends, family, school, and communities to use other energy sources: wind power, solar power, geothermal energy, and electric vehicles.
- Walk, bike, or use public transportation instead of going places by car.
- Turn off the lights when leaving a room.

OTHER WAYS TO HELP SEA TURTLES

Here are things you can do if you live near the ocean:
- If you see a stranded turtle on the beach, do *not* return it to the water. Instead, gently move it above the high-tide line, cover it with seaweed, mark the spot, and call your local office of NOAA (National Oceanographic and Atmospheric Administration).

- If you live near a turtle-nesting beach, darken your windows and turn off outside lights at night during egg-laying and hatching seasons. Newborn turtles are naturally drawn to light over water, so brighter lights on shore can lead them in the wrong direction. Shore lights can also discourage adult turtles from coming ashore to lay eggs.

Here are things you can do whether or not you live near the ocean:

- **Stop using plastic straws.** Every day people in the United States throw away millions of plastic straws. Many of these straws end up in rivers and oceans. Plastic straws can get stuck in turtles' noses so they can't breathe. Cities such as New York, Miami Beach, Seattle, and Washington, DC, have banned plastic straws. If plastic straws haven't been banned where you live, you can carry a reusable bamboo or metal straw.

- **Switch to reusable shopping bags.** Some cities and states—and even countries—have banned plastic bags. Still, millions of plastic bags end up in the ocean every year. They look just like jellyfish to a turtle but can block its digestive tract. Turtles who eat them are likely to die.

- **Don't buy helium balloons.** Loose balloons can land in the ocean, where turtles eat them. Balloons and their ribbons can both be deadly to turtles. One Kemp's ridley was found with a 16-foot ribbon tangled inside its body.

- **Become an activist.** Speak out! Even if you're not old enough to vote, you can write a postcard, make a phone call, or text or email the people who represent you in local government, Congress, or the White House. Tell them to take action on climate change. Encourage your friends to speak out, too.

- **Learn more about sea turtles and programs that protect them,** such as the New England Aquarium (http://www.neaq.org) and the Wellfleet Bay Wildlife Sanctuary (http://www.MassAudubon.org/).

- **Start a turtle fan club.** You might adopt a sea turtle through the Sea Turtle Conservancy (https://conserveturtles.org/). Talk about sea turtles with family and friends. Celebrate World Turtle Day every May 23.

Saving even one turtle makes a difference. After turtles mate, a female turtle will lay up to one hundred eggs each time she makes a nest. Not all the hatchlings will survive, but those that do will grow up, mate, and lay more eggs.

Turtles have been around since the time of the dinosaurs. Many people are working hard to make sure they continue to swim in our oceans.

Exactly where they belong.

A baby Kemp's ridley is the size of a half-dollar and weighs less than an ounce.

Acknowledgments

We would like to thank all the people who answered our many questions, welcomed us to their facilities, and shared their knowledge of Kemp's ridley sea turtles and their needs.

Mass Audubon Wellfleet Bay Wildlife Sanctuary:

Bob Prescott, founder of the Sea Turtle Rescue program

Christine Bates, community outreach coordinator

Jenette Kerr, marketing and communications coordinator

Karen Moore Dourdeville, marine biologist and sea turtle research coordinator

Melissa Lowe, regional director

Eamon Caffrey, Diamondback Terrapin Project coordinator

Jacey Corrente, sea turtle volunteer coordinator

Sasha Milsky, sea turtle rescue technician

Dylan Marat, seasonal field technician

Heather Pilchard, rescue volunteer

New England Aquarium's Sea Turtle Hospital:

Adam Kennedy, director of rescue and rehabilitation

Sarah Capozzoli, volunteer

Rebecca Lash, volunteer

New England Aquarium:

Pam Bechtold Snyder, director of marketing and communications

Suzanne Liola Matus, vice president, marketing, sales, and visitor experience

Kimberly Luciano, marketing and communications coordinator

Georgia Sea Turtle Center, Jekyll Island, Georgia:

Michelle Kaylor, director

Turtles Fly Too:

Ken Andrews, pilot and vice president

We are also grateful to other volunteers, on Cape Cod and at the Aquarium's hospital, who answered our questions as they worked to rescue, transport, and heal these endangered turtles.

Last but not least, thanks to John H. Straus for support, encouragement, and photos.

To the memory of Jenette Kerr, whose patient answers to our many questions and devotion to sea turtles and their survival helped to bring this book to life. And to all who love sea turtles, especially the many people who work to rescue, protect, and save them.
—L. K., P. R., J. B. M.

Photo Credits

Front cover and page 1: New England Aquarium. Pages 2–3: jesus/stock.adobe.com. Pages 4–5: Mass Audubon Wellfleet Bay. Pages 6–7: NASA. Page 8: Deb Felix. Page 9: Heather Pilchard. Pages 10–11: Mass Audubon Wellfleet Bay. Pages 12–13: John H. Straus. Page 14: Natalia Vázquez Torres. Pages 15: New England Aquarium. Pages 16–17: Mass Audubon Wellfleet Bay. Page 16 (inset): New England Aquarium. Pages 18–29: New England Aquarium. Page 27 (inset): gunillphotodesign/stock.adobe.com.

Pages 30–31: Ken Andrews. Page 31 (inset): New England Aquarium. Pages 32 and 33: Kristi Andrews. Pages 34–35: New England Aquarium. Pages 36–37: Ken Andrews. Pages 36 (inset) and 38–39: New England Aquarium. Pages 40–41: anita/stock.adobe.com. Pages 42–45: Natalia Vázquez Torres. Page 43 (inset): New England Aquarium. Pages 46–47: jesus/stock.adobe.com. Back cover (upper left): Heather Pilchard; (center and lower right): New England Aquarium.

At publication, all URLs in this book were accurate. Charlesbridge, the authors, and the photographers are not responsible for the content of any website.

Charlesbridge • 9 Galen Street Watertown, MA 02472
www.charlesbridge.com

Printed in China • OPIC
(hc) 10 9 8 7 6 5 4 3 2 1

Text type set in Italia
Edited by Natalia Vázquez Torres with assistance from Alyssa Mito Pusey
Designed by Diane M. Earley
Production supervised by Nicole Turner

Library of Congress Cataloging-in-Publication Data

Names: Ketchum, Liza, 1946– author. | Root, Phyllis, author. | Martin, Jacqueline Briggs, author.
Title: Turtles heading home / Liza Ketchum, Phyllis Root, and Jacqueline Briggs Martin.
Description: Watertown, MA: Charlesbridge, [2025] | Audience: Ages 6–10 | Audience: Grades 4–6 | Summary: "Join veterinarians and volunteers as they help rescue cold-stunned Kemp's ridley sea turtles, nurse them back to health, and release them in warmer waters."—Provided by publisher.
Identifiers: LCCN 2024012998 (print) | LCCN 2024012999 (ebook) | ISBN 9781623545864 (hardcover) | ISBN 9781632894564 (ebook)
Subjects: LCSH: Lepidochelys kempii—Juvenile literature. | Lepidochelys kempii—Conservation—Juvenile literature. | Lepidochelys—Juvenile literature. | Lepidochelys—Conservation—Juvenile literature.
Classification: LCC QL666.C536 K48 2025 (print) | LCC QL666.C536 (ebook) | DDC 597.92/8—dc23/eng/20240812
LC record available at https://lccn.loc.gov/2024012998
LC ebook record available at https://lccn.loc.gov/2024012999